Echoes in the Mind

Echoes in the Mind

Anthony Gilbert Rodriguez

CONTENTS

| VI | –

Introduction to Voice to Skull Technologies

Definition and Overview

Voice to Skull (V2K) technologies refer to a range of methods and devices designed to transmit auditory signals directly into a person's skull, bypassing the traditional pathways of sound. Initially developed for applications in military and law enforcement contexts, these technologies have garnered attention for their potential implications in various fields, including mental health, ethical considerations, and human rights. The ability to communicate through these methods raises significant questions about consent, the nature of reality, and the psychological effects experienced by those subjected to such technologies. This subchapter aims to provide a comprehensive overview of V2K technologies, exploring their definitions, functionalities, and the controversies surrounding them.

At the core of V2K technologies lies the principle of modulating sound waves to be perceived internally by the target individual. Various methods, such as microwave auditory effect and ultrasonic waves, can create auditory perceptions without the need for an external sound source. This phenomenon challenges conventional understandings of communication and consciousness, especially for those who believe in the existence of ethereal beings or alternate dimensions. The implications of V2K extend beyond mere technology, touching on the realms of spirituality and the human experience, where the lines between reality and perception can become blurred.

The ethical considerations surrounding V2K technologies are profound and multifaceted. Issues of consent, privacy, and psychological impact are central to the discourse. The potential misuse of these technologies raises alarms among civil liberties advocates, particularly concerning their application in law enforcement and surveillance. The capability to influence thoughts or behaviors covertly poses ethical dilemmas that demand rigorous scrutiny. For individuals who believe they are channeling ethereal beings, the intersection of technology and spirituality invites questions about the authenticity of their experiences and the potential for technological manipulation.

The psychological effects of V2K technologies are a growing concern, particularly regarding mental health and well-being. Reports from individuals who believe they have been targeted by V2K often include experiences of auditory hallucinations, paranoia, and anxiety. These effects can lead to significant distress and impact one's overall mental health. Understanding the psychological ramifications is crucial, especially in a society where mental health awareness is increasingly prioritized. The narratives of those affected must be approached with sensitivity, as they navigate the complexities of their experiences against the backdrop of technological realities.

As advancements in V2K technologies continue to unfold, the legal frameworks and regulations governing their use remain inadequately defined. The lack of clear guidelines creates a landscape ripe for exploitation and misuse, further complicating public perception and contributing to misinformation. Case studies of alleged V2K technology usage highlight the need for transparency and accountability in both military and civilian contexts. As society grapples with the implications of these technologies, it is essential to foster informed discussions that bridge the gap between technological development and ethical responsibility, ensuring that the rights and dignity of individuals are upheld in an increasingly complex world.

Historical Background

The historical background of Voice to Skull technologies (V2K) traces its roots to the intersection of communication theory, military research, and psychological manipulation. Emerging from early experiments in psychophysics during the mid-20th century, researchers began to explore the potential of transmitting sound directly into the human auditory system without conventional means. Initial investigations were largely focused on understanding human perception and the effects of sound on behavior. By the 1970s, advancements in neuroscience and technology led to more sophisticated attempts at manipulating sound frequencies, which were later adapted for military and intelligence applications.

In the context of military usage, V2K technologies gained traction during the Cold War, as governments sought innovative methods for psychological warfare and enhanced interrogation techniques. The technological capabilities developed during this period laid the groundwork for various covert operations that aimed to disrupt enemy morale and influence decision-making processes. It is during this era that the ethical implications of such technologies began to surface, raising concerns about the potential for abuse and the violation of human rights. The dual-use nature of these technologies, capable of both beneficial and harmful applications, has fueled ongoing debates regarding their moral standing.

The psychological effects of V2K are critical to understanding its implications. Individuals subjected to these technologies often report experiences akin to auditory hallucinations, which can lead to severe mental health issues, including anxiety, paranoia, and depression. These psychological impacts can be compounded by the stigma surrounding mental illness, causing victims to feel isolated and misunderstood. The overlap between V2K experiences and reported encounters with ethereal beings adds another layer of complexity, as some individuals interpret their experiences through the lens of spiritual or extraterrestrial phenomena, further blurring the lines between reality and perception.

As V2K technologies have evolved, their potential applications in law enforcement and surveillance have raised significant ethical concerns. The ability to influence thoughts and behaviors covertly poses threats not only to individual freedoms but also to democratic institutions. Legal frameworks governing the use of such technologies remain underdeveloped, leading to a patchwork of regulations that vary by jurisdiction. Public perception of V2K is often clouded by misinformation, conspiracy theories, and sensational media portrayals, which complicate meaningful discourse about the ethical use of these technologies and their implications for society.

The future of Voice to Skull technologies is fraught with both potential advancements and ethical dilemmas. As technological capabilities continue to improve, the implications for mental health, personal autonomy, and human rights will only grow more pronounced. Case studies of alleged V2K usage often reveal a disturbing pattern of misuse, highlighting the urgent need for comprehensive regulations and public education. Understanding the historical context of these technologies is essential for navigating the complex interplay between innovation, ethics, and the human experience in an increasingly interconnected world.

Purpose and Scope of the Book

This book aims to explore the intricate and often controversial realm of Voice to Skull (V2K) technologies, shedding light on both their implications and the broader ethical considerations surrounding their use. In addressing an audience composed of conspiracy theorists, spiritual and extraterrestrial channelers, as well as those who believe they are receiving messages from ethereal beings, this text seeks to bridge the gap between scientific inquiry and the metaphysical interpretations of such technologies. By delving into the purpose and scope of this book, readers will gain a comprehensive understanding of how V2K technologies intersect with various aspects of human experience, belief systems, and societal structures.

The implications of Voice to Skull technologies extend far beyond mere technical applications; they touch upon fundamental questions of human rights, mental well-being, and the ethical frameworks that govern their use. The exploration of these technologies encompasses their applications in law enforcement and military settings, where they can potentially alter the landscape of surveillance and control. By examining the psychological effects on individuals who may perceive themselves as targets, the book will provide valuable insights into the intersection of technology, psychology, and personal experience, allowing readers to contextualize their beliefs within a larger narrative.

In addition to discussing the psychological ramifications, this text will address the impact of V2K technologies on mental health and well-being. The relationship between perceived external influences and mental health challenges is a crucial aspect of this discourse. By emphasizing case studies of alleged V2K usage, the book aims to illustrate real-world applications and the resultant experiences of individuals, thus enriching the conversation around the subject. This examination will also serve to demystify some of the misinformation that surrounds these technologies, fostering a more informed discourse among those who are deeply invested in the metaphysical aspects of V2K.

Furthermore, the legal frameworks and regulations surrounding Voice to Skull technologies will be critically assessed. Understanding the current laws and policies governing these technologies is essential for any meaningful dialogue about their ethical use. This book will highlight the ongoing debates surrounding privacy rights and the potential for misuse, offering a balanced perspective that encompasses both the fears and the potential benefits associated with V2K technologies. Through this analysis, readers will be encouraged to reflect on their own beliefs and the societal implications of these emerging technologies.

Finally, the book will explore technological advances and future developments in Voice to Skull technologies, considering their potential applications in various fields. As these technologies evolve, so too must our understanding of their implications on human rights and ethical considerations. By engaging with this material, readers will not only gain insights into V2K technologies but also engage in a broader conversation about the nature of communication, consciousness, and the unseen forces that may be at play in their lives. Through this exploration, the book aims to empower readers to navigate the complexities of V2K technologies with a discerning and informed mindset.

Implications of Voice to Skull Technologies

Societal Impact

The emergence of Voice to Skull (V2K) technologies has ignited significant discourse regarding their societal impact. These technologies, which enable the transmission of sound directly into an individual's skull, raise critical questions about ethical considerations and the potential for misuse. As society grapples with the implications of such advancements, it becomes essential to explore the boundaries of acceptable use, especially when considering the intersection of technology and human rights. The potential for surveillance and manipulation, particularly in the realms of law enforcement and military applications, underscores the need for a robust legal framework to protect individuals from invasive practices.

The ethical ramifications of V2K technologies extend beyond mere surveillance; they delve into the core of personal autonomy and mental integrity. The ability to implant thoughts or influence behaviors without consent poses profound moral dilemmas. For people who believe they are channeling ethereal beings or those attuned to spiritual dimensions, the potential for confusion between external technological influence and genuine spiritual communication can be disorienting. This blurring of lines challenges existing ethical norms and calls for a reevaluation of how society defines and safeguards free will in an age where technology can easily manipulate perception.

Another crucial aspect of V2K technologies is their psychological impact on individuals subjected to such experiences. The psychological effects can range from heightened anxiety and paranoia to a distorted sense of reality. Those who feel they are being targeted may experience significant distress, leading to negative implications for their mental health and overall well-being. It is vital for mental health professionals to recognize the influence of these technologies on individuals' psychological states and to offer support that acknowledges both the technological and the experiential dimensions of their suffering.

The role of V2K technologies in military applications and law enforcement raises additional concerns about their societal implications. While proponents may argue that these technologies enhance operational capabilities, the potential for abuse is alarming. The use of V2K in interrogation tactics or surveillance operations poses ethical questions about the treatment of individuals and the extent to which governments can exert control over their citizens. Additionally, the lack of transparency surrounding the deployment of such technologies can lead to widespread mistrust in institutions, further fracturing the relationship between the public and authorities.

Finally, as society navigates the complexities of V2K technologies, public perception plays a significant role in shaping discourse and policy. Misinformation and conspiracy theories often cloud the understanding of these technologies, leading to polarized views that hinder constructive dialogue. Educating the public about the realities of V2K technologies, their potential applications, and the ethical concerns they raise is crucial. By fostering an informed community, individuals can engage in meaningful discussions about the future of these technologies, advocating for responsible regulation that prioritizes human rights and dignity in an increasingly interconnected world.

Ethical Dilemmas

Ethical dilemmas surrounding Voice to Skull (V2K) technologies are multifaceted, intertwining issues of consent, privacy, and the potential for misuse. As the technology evolves, the implications for individual autonomy become increasingly concerning. The ability to transmit voices or messages directly into a person's mind raises questions about the fundamental right to control one's own thoughts and perceptions. For those who are channeling ethereal beings or believe in spiritual communication, the possibility of interference from such technologies presents a profound dilemma. How can one discern between genuine spiritual guidance and artificially induced voices, leading to confusion and potential mental distress?

The ethical considerations in the development and deployment of V2K technologies are particularly pressing in contexts such as law enforcement and military applications. While proponents argue that these tools can enhance security measures, the potential for abuse is significant. The prospect of using V2K for surveillance or interrogation purposes raises alarms about human rights violations and the erosion of civil liberties. Those who subscribe to conspiracy theories may view these technologies as tools of oppression, further exacerbating distrust in governmental and institutional authorities. The line between protection and invasion becomes blurred, and ethical oversight becomes paramount to prevent misuse.

Psychological effects are another critical aspect of the ethical landscape surrounding V2K technologies. The experiences of individuals who claim to be targeted by these technologies can lead to severe mental health repercussions, including anxiety, paranoia, and delusions. This raises ethical questions about the responsibility of developers and users to consider the psychological impact on individuals. For those who perceive themselves as channeling ethereal beings, the intrusion of V2K could distort their spiritual experiences, leading to existential crises or a breakdown in their understanding of reality. A framework for ethical responsibility must be established to protect vulnerable populations from the unintended consequences of these technologies.

Moreover, the legal framework and regulations governing V2K technologies are still in their infancy. As the technology becomes more pervasive, there is an urgent need for comprehensive guidelines that address ethical considerations. Current laws may not adequately protect individuals from the potential infringements associated with V2K applications. For conspiracy theorists and those engaged in spiritual practices, the lack of legal clarity can foster an environment rife with misinformation and fear. The challenge lies in creating regulations that balance innovation with ethical standards while ensuring the protection of human rights.

Public perception plays a crucial role in shaping the discourse around V2K technologies. Misinformation can lead to widespread fear and misunderstanding, complicating efforts to address the ethical dilemmas they present. For those who believe they are channeling higher entities, the reality of V2K could lead to skepticism about their experiences, potentially undermining their spiritual beliefs. It is imperative to foster informed discussions that separate fact from fiction, allowing for a deeper understanding of how V2K technologies intersect with personal beliefs, ethical considerations, and societal implications. As we navigate the future of these technologies, ongoing dialogue about their ethical ramifications will be essential to ensure they serve humanity positively.

Technological Dependency

Technological dependency has become a defining characteristic of modern society, shaping the way individuals interact with the world and with each other. In the context of Voice to Skull technologies, this dependency raises significant questions about the implications for personal autonomy and mental health. As individuals increasingly rely on technology for communication and information processing, the potential for these technologies to manipulate thoughts and perceptions presents a unique challenge. It is crucial to explore how this dependency might affect individual agency, especially for those who believe they are channeling ethereal beings or engaging with spiritual dimensions.

The ethical considerations surrounding Voice to Skull technologies are deeply intertwined with the issue of technological dependency. As people become more accustomed to technology that can influence their thoughts, there is a risk of normalizing invasive practices that could undermine personal freedom. For those who engage with these technologies under the belief that they are connecting with higher consciousness or extraterrestrial entities, the ethical implications are even more pronounced. The line between voluntary engagement and coercion becomes blurred, raising concerns about informed consent and the morality of using such technologies on vulnerable populations.

Psychological effects are another critical aspect of the dependency on Voice to Skull technologies. Many individuals report experiencing heightened anxiety, paranoia, and confusion, which may stem from the constant barrage of auditory stimuli that these technologies can produce. The psychological impact can lead to a deterioration of mental health, particularly for those already predisposed to mental illness or who are susceptible to suggestion. For conspiracy theorists and spiritual channelers, this can create a feedback loop where the perceived influence of these technologies reinforces their beliefs, complicating their relationship with reality and further entrenching their dependency.

In law enforcement and surveillance, the application of Voice to Skull technologies raises significant ethical and legal questions. As authorities adopt these tools for monitoring and control, the potential for abuse increases, particularly concerning civil liberties and human rights. The dependency on such technologies can lead to a culture of surveillance that undermines trust within communities. Those who believe they are experiencing targeted harassment through these technologies may find themselves in a psychological battle that challenges their perceptions of reality, prompting further examination of the societal implications of widespread technological reliance.

Looking towards the future, the technological advances in Voice to Skull technologies promise to deepen this dependency, with potential applications in military, law enforcement, and even therapeutic settings. However, as these technologies evolve, so too must the legal frameworks and regulations that govern their use. The challenge lies in balancing innovation with ethical considerations and protecting individual rights. Public perception, often clouded by misinformation, will heavily influence the acceptance and regulation of these technologies. For those who channel ethereal beings or engage with spiritual realms, the implications of such dependency could lead to a profound re-examination of the nature of consciousness itself, urging a collective inquiry into the boundaries of technology and the human experience.

Ethical Considerations in Voice to Skull Technolog

Informed Consent

Informed consent is a foundational concept in the ethical landscape surrounding voice to skull technologies, particularly given their controversial applications and implications. This principle mandates that individuals must be fully aware of, and agree to, the procedures and potential risks before any intervention occurs. In the context of voice to skull technologies—where individuals report experiences of hearing voices or receiving messages through non-traditional means—the need for informed consent becomes even more critical. Many users may be unaware of the implications of these technologies on their mental and emotional well-being, thereby raising ethical concerns about how consent is obtained and understood.

The ethical considerations tied to informed consent in relation to voice to skull technologies extend beyond mere acknowledgment. They delve into the nuances of understanding what is being consented to, especially in cases where individuals may already believe they are channeling ethereal beings or experiencing otherworldly communication. This belief system can complicate the process of informed consent, as individuals may interpret their experiences through a spiritual lens, making it challenging to delineate between voluntary participation and coercion or manipulation. The blurred lines can lead to ethical dilemmas, especially when the technology is employed in environments such as law enforcement or military applications.

Psychologically, the implications of voice to skull technologies on informed consent cannot be overlooked. Individuals who believe they are receiving messages from external sources may not fully grasp the nature of their experiences, leading to a disconnect between their understanding and the realities of the technology. This disconnect can result in significant mental health challenges, such as anxiety, paranoia, or depression, particularly if individuals feel they lack control over their experiences. The psychological impact of these technologies necessitates a robust informed consent process that includes comprehensive education about the potential effects and nature of the technology involved.

In the realm of law enforcement and surveillance, informed consent takes on an even more complex dimension. The use of voice to skull technologies in these sectors raises questions about the legality and morality of employing such methods without explicit consent from the individuals involved. The potential for misuse or abuse of these technologies is substantial, particularly when individuals are subjected to surveillance or interrogation without their full knowledge. Establishing a legal framework that prioritizes informed consent is essential to protect individual rights and ensure ethical practices in these sensitive areas.

The public perception and misinformation surrounding voice to skull technologies further complicate the informed consent landscape. Many individuals may hold misconceptions about how these technologies operate or the extent to which they are used, leading to skepticism or fear. To foster a more informed populace, it is crucial to provide accurate information and engage in open dialogue about the ethical implications, psychological effects, and legal regulations. This effort will not only empower individuals to make informed choices but also contribute to a more nuanced understanding of the technologies that are increasingly becoming part of contemporary society.

Privacy Violations

Privacy violations in the realm of voice to skull technologies present a profound concern that extends beyond individual rights to encompass broader societal implications. As these technologies enable direct communication with the mind, they inadvertently blur the lines between public and private spheres. Individuals may find their thoughts, emotions, and even personal experiences susceptible to external influence or surveillance, raising questions about autonomy and consent. The ramifications of such invasions can lead to a pervasive sense of vulnerability, as anyone could potentially be subjected to unwarranted psychological intrusions without their knowledge or agreement.

The ethical considerations surrounding the use of voice to skull technologies are paramount in discussions of privacy. The potential for misuse by governments or private entities poses a significant threat to civil liberties. The ability to implant thoughts or manipulate perceptions raises alarms about the power dynamics involved, particularly when targeting marginalized groups. The ethical dilemma intensifies when considering the implications for informed consent; individuals may unwittingly become subjects of experimentation or manipulation without the knowledge of the technology's existence or its intended use. This highlights the urgent need for ethical guidelines and frameworks that prioritize human rights and individual dignity.

Psychological effects stemming from privacy violations linked to voice to skull technologies are profound. Individuals who believe they are being subjected to such technologies may experience heightened anxiety, paranoia, and a distorted sense of reality. These psychological ramifications can have devastating impacts on mental health, as the constant fear of surveillance and mind control can lead to social withdrawal and a breakdown of trust in relationships. Additionally, the stigma associated with these experiences can further isolate individuals, complicating their efforts to seek help or validation from the broader community.

In the context of law enforcement and surveillance, the implications of voice to skull technologies extend to a potential erosion of privacy rights. The use of these technologies by authorities can create a chilling effect on free expression and dissent. When citizens fear that their thoughts may be monitored or influenced, the natural discourse essential to a democratic society is stifled. This raises significant legal questions, as existing frameworks may not adequately address the unique challenges posed by such invasive technologies. The need for comprehensive legislation to protect individuals from unwarranted surveillance and psychological manipulation is becoming increasingly urgent.

The public perception of voice to skull technologies often intertwines with misinformation and conspiracy theories, complicating the dialogue around privacy violations. While some believe these technologies are merely the stuff of science fiction, others assert that they are already in use, leading to a polarized understanding of their implications. This division can hinder meaningful conversations about ethical guidelines, legal protections, and mental health resources for those affected. As technological advancements continue to progress, fostering an informed and nuanced public discourse will be essential in addressing the multifaceted challenges posed by voice to skull technologies and their potential to infringe upon personal privacy.

Moral Responsibility

Moral responsibility in the context of Voice to Skull (V2K) technologies is a complex issue that intertwines ethical considerations, psychological impacts, and implications for human rights. As the potential applications of V2K technologies expand, so does the responsibility of those who develop, implement, and regulate these systems. It is crucial for innovators and policymakers to reflect on the moral implications of their actions, as their decisions can significantly affect individual autonomy and mental well-being. The ethical dilemmas surrounding the use of V2K technologies necessitate a robust framework that prioritizes human dignity and considers the voices of those who may be subjected to these technologies without their consent.

The psychological effects of V2K technologies warrant particular attention, especially regarding individuals who believe they are experiencing these phenomena. Those who perceive themselves as recipients of messages through V2K may grapple with feelings of paranoia, anxiety, and isolation. It is essential to acknowledge the mental health implications that arise when technology intersects with personal reality. The responsibility lies not only with the creators of such technologies but also with the community at large to provide support and understanding to those affected. Acknowledging these psychological burdens can foster a more compassionate dialogue about the implications of V2K technology on mental health and well-being.

When considering the use of V2K technologies in law enforcement and surveillance, moral responsibility takes on an even greater significance. The potential for misuse, whether intentional or accidental, raises questions about civil liberties and accountability. Law enforcement agencies must navigate the fine line between ensuring public safety and respecting individual rights. The ethical application of V2K technologies requires transparency and strict regulations to prevent abuse and to protect vulnerable populations. A commitment to ethical conduct in this area is essential to maintain public trust and uphold the values of justice and fairness.

Military applications of V2K technologies present unique ethical challenges. The use of such technologies in warfare or psychological operations raises critical questions about the moral implications of manipulating human perception and behavior. As military organizations explore the potential of V2K for tactical advantages, the need for ethical guidelines becomes paramount. The international community must engage in discussions regarding the acceptable use of these technologies to ensure that they do not infringe upon human rights or dignity. The moral responsibility extends beyond national borders, emphasizing the need for global standards governing the deployment of V2K technologies in conflict situations.

Public perception and misinformation surrounding V2K technologies also play a vital role in the ongoing discourse about moral responsibility. The spread of conspiracy theories can lead to fear and misunderstanding, complicating the conversation about the ethical use of these technologies. It is crucial for credible voices to emerge and provide accurate information to counteract misinformation. Engaging with communities that feel marginalized or targeted by V2K technologies requires sensitivity and openness. By fostering a well-informed public, we can work towards a more nuanced understanding of the ethical implications and responsibilities associated with the development and application of Voice to Skull technologies.

IV

Psychological Effects of Voice to Skull Technologi

Perception of Reality

Perception of reality is a complex tapestry woven from individual experiences, beliefs, and external stimuli. In the context of Voice to Skull (V2K) technologies, this perception is profoundly influenced by the interplay between perceived external voices and internal cognitive processes. For individuals who believe they are channeling ethereal beings or receiving messages from higher realms, the introduction of V2K technologies can blur the line between genuine spiritual communication and artificially induced auditory experiences. This phenomenon raises critical questions about the nature of reality: Are these communications truly from otherworldly sources, or are they the product of technology manipulating the mind?

The ethical implications of V2K technologies further complicate this perception. The potential for misuse in surveillance and control can lead to a sense of paranoia and distrust in one's environment, which can warp an individual's understanding of reality. When people are subjected to perceived external influences without their consent, it challenges their autonomy and self-perception. This ethical dilemma is particularly poignant for those who engage with spiritual practices, as it raises concerns about the authenticity of their experiences and whether they are genuinely connecting with higher consciousness or being subjected to technological manipulation.

Psychologically, the impact of V2K technologies on perception is significant. Individuals may experience heightened anxiety or confusion as they grapple with conflicting narratives about their experiences. For those who believe they are channeling messages, the introduction of V2K can lead to a fracturing of their perceived reality, as they may question the validity of their insights. This psychological tension can result in a range of mental health issues, including paranoia, anxiety disorders, and even psychosis, as individuals struggle to discern between their internal thoughts and external influences.

In law enforcement and military applications, the use of V2K technologies raises profound implications for societal perception. While these technologies may be employed for security and control, they also present a chilling possibility of widespread surveillance that can alter the fabric of personal reality. When individuals become aware of the potential for technology to invade their thoughts, their trust in institutions may erode, leading to a broader societal skepticism. This shift in perception can create a divide where individuals either embrace or reject the narratives surrounding their experiences, further complicating public discourse.

Ultimately, the public perception and misinformation surrounding V2K technologies contribute significantly to the overall understanding of reality. As narratives proliferate across various platforms, they shape collective beliefs about what is possible and what is merely a figment of imagination. For conspiracy theorists and those engaged in spiritual practices, the challenge lies in discerning truth from fiction in an age where technology can convincingly mimic the voices of the unseen. The implications of these technologies extend far beyond individual experience, touching upon human rights, ethical considerations, and the very essence of what it means to perceive reality in a technologically advanced society.

Anxiety and Paranoia

Anxiety and paranoia represent significant psychological responses that can arise in individuals exposed to the implications of Voice to Skull technologies. These technologies, often discussed within conspiracy circles, involve the transmission of auditory signals directly into the human mind, bypassing traditional sound pathways. For those who believe they are channeling messages from ethereal beings, the experience may be compounded by a heightened sensitivity to perceived external influences, leading to a distorted perception of reality. In such contexts, the line between genuine spiritual experiences and induced psychological distress can blur, resulting in profound anxiety about one's mental state and the origins of perceived communications.

The ethical considerations surrounding Voice to Skull technologies are paramount, particularly as they relate to the mental health of individuals. The potential for misuse of these technologies raises pressing questions about consent, autonomy, and the right to mental privacy. Those who experience unsolicited auditory stimuli may find themselves grappling with feelings of helplessness and fear, as they navigate a world where their thoughts and perceptions could be manipulated or influenced by unseen forces. The ethical implications thus extend beyond individual distress, impacting broader societal norms regarding mental health care and the responsibilities of those who develop and deploy such technologies.

Psychologically, the effects of Voice to Skull technologies can manifest as heightened anxiety and paranoia, particularly in individuals predisposed to such conditions. The experience of hearing voices or receiving messages can lead to increased isolation, as individuals may withdraw from social interactions out of fear of misunderstanding or ridicule. This withdrawal can exacerbate feelings of paranoia, as the individual becomes increasingly convinced that they are being targeted or monitored. The cyclical nature of these psychological effects underscores the need for comprehensive mental health support for those who believe they are affected by these technologies, as well as greater awareness and understanding within the broader community.

In the context of law enforcement and surveillance, the potential use of Voice to Skull technologies raises significant concerns about privacy and mental health. The fear that one's thoughts could be intercepted or manipulated by authorities contributes to a pervasive sense of paranoia among individuals aware of these technologies. This anxiety can hinder trust in institutions and exacerbate societal divisions, as those who feel targeted by such surveillance may cultivate a deep-seated mistrust of law enforcement. The impact on mental health is profound, as individuals navigate the dual challenges of real-world interactions and the perceived intrusions of technology into their private mental spaces.

As discussions around Voice to Skull technologies continue to evolve, it is crucial to address the mental health implications associated with their use. Individuals who believe they are experiencing these technologies often report feelings of anxiety and paranoia that can significantly impair their quality of life. Acknowledging these psychological effects is essential for fostering a supportive environment where individuals can seek help and validation without fear of stigma. Moving forward, a concerted effort to educate the public about the realities and myths surrounding Voice to Skull technologies will be vital in alleviating unnecessary anxiety and promoting mental well-being amidst the backdrop of rapidly advancing technology.

Identity and Self-Perception

Identity and self-perception are deeply interwoven with the experiences individuals have in relation to voice to skull technologies. For those who believe they are channeling ethereal beings or receiving messages from other dimensions, the intrusion of external voices can significantly alter their understanding of self. The auditory stimuli perceived as originating from these technologies can lead individuals to question the authenticity of their thoughts and beliefs. This creates a complex interplay between the individual's sense of self, their perceived reality, and the external influences they believe are at play. As such, identity becomes malleable, shaped by interactions with what they interpret as otherworldly communications.

The psychological effects of voice to skull technologies extend beyond mere auditory hallucinations. Many individuals report feelings of paranoia, anxiety, or even a sense of purpose linked to the messages they receive. This can lead to an altered self-perception, where individuals may view themselves as conduits for higher knowledge or as targets of malevolent forces. The implications of these experiences can be profound, affecting how individuals relate to others and navigate their daily lives. The constant questioning of one's mental state can create an internal conflict, as the distinction between the self and the perceived external influence blurs.

Moreover, the ethical considerations surrounding voice to skull technologies are critical in understanding their impact on identity. The potential for misuse in law enforcement or military applications raises concerns about the violation of individual rights and the manipulation of personal beliefs. When individuals are subjected to external voices, the autonomy of their thoughts and identity comes into question. This ethical dilemma not only affects those targeted by such technologies but also reflects broader societal implications regarding consent and the fundamental right to mental privacy. The ability to influence or disrupt an individual's self-perception through technology poses significant moral challenges that society must confront.

Public perception of voice to skull technologies often oscillates between skepticism and fear. Misinformation can exacerbate feelings of isolation among those who genuinely believe they are experiencing these phenomena. This distorted public view can impact how individuals identify themselves within broader communities, either as victims of technological manipulation or as enlightened beings receiving messages from beyond. The challenge lies in fostering a more nuanced understanding that acknowledges both the potential for technological abuse and the genuine psychological experiences of individuals. A balanced approach is necessary to help individuals maintain a coherent sense of identity amidst the noise of external voices.

As technology continues to advance, the future developments in voice to skull technologies will likely raise further questions about identity and self-perception. The potential for enhanced capabilities may deepen the psychological impact on those who perceive these communications as real. Continuous dialogue about the implications of such technologies, particularly regarding mental health and human rights, is essential. By addressing these concerns, we can work towards a society that respects individual identity while navigating the complex landscape of emerging technologies and their effects on the human experience.

V

Voice to Skull Technologies in Law Enforcement and

Utilization in Crime Prevention

Voice to Skull technologies have emerged as a controversial tool in the realm of crime prevention, raising significant questions about their ethical implications and potential psychological effects on individuals. Law enforcement agencies have been exploring these technologies as a means to enhance surveillance capabilities and deter criminal activities. By utilizing directed sound waves that can transmit messages directly into a person's mind, authorities claim they can communicate warnings or commands to suspects without alerting others, potentially preventing imminent crimes. This capability has sparked considerable debate regarding the balance between public safety and individual rights.

The psychological effects of employing Voice to Skull technologies in crime prevention cannot be understated. Individuals subjected to these transmissions may experience heightened anxiety, paranoia, or disorientation, which can lead to unintended consequences. A suspect may react unpredictably when they perceive themselves as being monitored or influenced by an unseen force. This raises ethical concerns about the mental well-being of individuals targeted by such technologies, particularly in high-stress situations where their behavior could be misinterpreted as threatening or aggressive.

Furthermore, the implications for human rights are profound. The use of Voice to Skull technologies in law enforcement must navigate the delicate line between ensuring public safety and upholding constitutional rights. Critics argue that the deployment of such technologies infringes upon the right to privacy and the right to mental autonomy. The potential for misuse or overreach by authorities poses a significant risk, leading to calls for stringent legal frameworks and regulations that govern the use of these technologies in crime prevention efforts.

Public perception plays a critical role in the ongoing discourse surrounding Voice to Skull technologies. Misinformation and conspiracy theories can cloud understanding and fuel fears about their applications. Many who believe in the existence of these technologies often view them through the lens of personal experience or anecdotal evidence, which can complicate the dialogue about their legitimate use in crime prevention. Educating the public about the technology's capabilities, limitations, and ethical considerations is essential to fostering a more informed debate.

As technological advancements continue to evolve, the future of Voice to Skull technologies in crime prevention remains uncertain. While they hold potential for innovative approaches to policing, careful consideration of their psychological impact, ethical ramifications, and legal boundaries is imperative. Ongoing research, case studies, and open discussions will be crucial in determining how these technologies can be utilized responsibly, ensuring that the pursuit of safety does not come at the expense of individual rights and mental health.

Controversies and Backlash

The emergence of Voice to Skull (V2K) technologies has sparked significant controversy and backlash, particularly among those who perceive these advancements as encroachments on human rights and personal freedoms. Many conspiracy theorists and advocates for spiritual and extraterrestrial communication view V2K as a tool of manipulation, suggesting that it might be used to infiltrate the minds of individuals, creating a sense of paranoia and distrust among the populace. This perception is compounded by reports of alleged governmental and military applications of these technologies, with claims that they are utilized for psychological warfare, surveillance, and control over dissidents. Such allegations highlight the ethical dilemmas surrounding the deployment of V2K, raising questions about consent, autonomy, and the potential for abuse by those in power.

The ethical implications of Voice to Skull technologies are a focal point of contention. Critics argue that the ability to transmit thoughts or commands directly to an individual's mind infringes upon the fundamental right to privacy and self-determination. This concern is particularly pronounced within communities that emphasize the spiritual and metaphysical aspects of existence, where the sanctity of the mind is often viewed as a sacred space. Those who believe in channeling ethereal beings posit that V2K technologies could disrupt or distort genuine spiritual experiences, leading to confusion and disillusionment among seekers of truth. As these technologies evolve, the ethical considerations must grapple with the balance between potential benefits, such as therapeutic applications, and the risks of psychological manipulation and coercion.

The psychological effects of Voice to Skull technologies are another area of significant concern, particularly regarding mental health and well-being. Individuals who believe they have been subjected to V2K often report feelings of anxiety, depression, and a loss of control over their thoughts. This phenomenon has led to a growing discourse around the impact of V2K on vulnerable populations, including those with pre-existing mental health conditions. The interplay between technology and mental health raises fundamental questions about responsibility and accountability, particularly when individuals are led to believe their thoughts are being externally influenced. As mental health professionals and researchers delve into the implications of V2K technologies, understanding the psychological toll on individuals remains paramount.

The use of Voice to Skull technologies in law enforcement and surveillance has also incited backlash, with fears that such tools could facilitate invasive policing practices. Critics argue that the potential for abuse is high, particularly in a landscape where civil liberties are already under threat. The idea that law enforcement agencies might use V2K to control or intimidate individuals raises profound concerns about the legality and morality of such actions. The lack of clear regulations governing the use of these technologies further exacerbates the situation, leaving room for speculation and distrust among the public. As discussions surrounding V2K technologies continue, the call for a robust legal framework becomes increasingly urgent to protect citizens from potential misuse.

Public perception of Voice to Skull technologies is heavily influenced by misinformation and sensationalism, which can further entrench the divide between believers and skeptics. Many narratives surrounding V2K are steeped in conspiracy theories, which often overshadow legitimate concerns and hinder constructive dialogue. The challenge lies in navigating this complicated landscape, where fear and suspicion can cloud judgment and lead to a backlash against technological innovation. As new developments arise in V2K research, it is essential to foster an informed public discourse that addresses both the potential benefits and risks associated with these technologies. Only through open and honest conversations can society hope to untangle the complex web of implications surrounding Voice to Skull technologies and their impact on personal and collective consciousness.

Balancing Security and Privacy

In the realm of voice to skull technologies, the balance between security and privacy is a contentious issue that evokes strong feelings among various communities. Proponents argue that these technologies can enhance security measures, particularly in law enforcement and military applications, providing a tool for efficient communication and threat detection. However, the implications for personal privacy are profound, raising ethical questions about consent and the potential for misuse. As these technologies become more sophisticated, the challenge lies in ensuring that security measures do not infringe upon individual privacy rights, leading to a societal environment where surveillance becomes the norm.

The ethical considerations surrounding voice to skull technologies are multifaceted. On one hand, there is a legitimate need for security in certain contexts, such as preventing crime or terrorism. On the other hand, the deployment of such technologies without clear regulations can lead to violations of civil liberties. For individuals who believe they are channeling ethereal beings or experiencing heightened states of consciousness, the intrusion of external voices, whether real or perceived, can disrupt their mental and spiritual practices. The ethical responsibility of developers and authorities must extend beyond mere functionality to encompass respect for individuals' rights to privacy and mental integrity.

Psychologically, the impact of voice to skull technologies on mental health and well-being is significant. For those who are sensitive to auditory stimuli or who are already vulnerable, the experience of having voices transmitted into their minds can lead to feelings of paranoia, anxiety, and distress. This psychological burden is compounded when individuals feel they are being surveilled or manipulated, leading to a cycle of fear and mistrust. Understanding these psychological effects is crucial for establishing frameworks that protect individuals while allowing for the legitimate use of such technologies in security contexts.

In the realm of law enforcement and surveillance, the use of voice to skull technologies raises important questions about accountability and oversight. While these technologies can provide law enforcement with enhanced capabilities, the potential for abuse is substantial. Instances of alleged misuse have fueled public skepticism and conspiracy theories, particularly among those who believe in the potential for government overreach. Establishing a legal framework that governs the use of these technologies is essential to ensuring transparency and protecting citizens from potential violations of their rights.

As our society grapples with the rapid advancements in voice to skull technologies, it is imperative to engage in open dialogues about their implications for security and privacy. Public perception often skews towards fear and misinformation, complicating the discourse around these technologies. By fostering an informed discussion that acknowledges both the potential benefits and risks, we can work towards a balanced approach that respects individual rights while addressing legitimate security concerns. This balance is essential not only for the integrity of democratic societies but also for the mental and spiritual well-being of individuals navigating the complexities of modern technological landscapes.

VI

Impact on Mental Health and Well-being

Effects on Vulnerable Populations

The emergence of Voice to Skull (V2K) technologies has raised significant concerns regarding their impact on vulnerable populations, including individuals with mental health challenges, those experiencing homelessness, and marginalized communities. These groups often lack the resources, support systems, and platforms to voice their experiences and concerns about the potential misuse of such technologies. Vulnerable populations may be disproportionately affected by V2K applications, leading to exacerbated feelings of isolation, anxiety, and paranoia as they grapple with the implications of voices perceived as external intrusions. The psychological burden of believing one is being targeted by such technologies can further alienate these individuals from society, creating a cycle of distress that is difficult to break.

Ethical considerations surrounding V2K technologies are paramount when examining their effects on vulnerable populations. The potential for misuse by authorities, including law enforcement and military applications, raises questions about consent, privacy, and the autonomy of individuals who may not fully understand the technologies or their implications. The use of V2K technologies in surveillance and control can lead to a chilling effect on personal freedoms, particularly for those already marginalized. Ethical frameworks must be established to protect vulnerable individuals from exploitation and to ensure that their rights are upheld in the face of rapidly advancing technology.

The psychological effects of V2K technologies on vulnerable populations can be profound and multifaceted. Individuals who perceive themselves as targets of such technologies may experience heightened anxiety, depression, and a distorted sense of reality. The belief that one is being monitored or manipulated can lead to a deep mistrust of others and may result in withdrawal from social interactions. Furthermore, the stigma associated with mental health issues can be compounded by the belief in V2K technologies, leading to further isolation and a lack of understanding from those around them. This creates a feedback loop that can deteriorate mental health, making intervention and support crucial.

Legal frameworks and regulations regarding V2K technologies are still in their infancy, and this lack of oversight has significant implications for vulnerable populations. The absence of clear guidelines leaves individuals open to potential abuse and exploitation without recourse. Vulnerable groups may find themselves at the mercy of ambiguous laws that do not adequately protect their rights or address their unique challenges. Advocacy for stronger legal protections and a comprehensive understanding of the implications of V2K technologies is essential to safeguard the interests of those most at risk.

Public perception and misinformation about V2K technologies further complicate the issue for vulnerable populations. Many individuals may dismiss claims regarding the existence or effects of these technologies, leading to a lack of empathy and understanding for those who report experiencing them. This disconnect can leave vulnerable individuals feeling unheard and invalidated, further exacerbating their mental health struggles. Education and awareness campaigns are critical to foster a more informed public discourse around V2K technologies, ensuring that the experiences of all individuals, particularly those in vulnerable positions, are recognized and addressed appropriately.

Therapeutic Interventions

Therapeutic interventions related to Voice to Skull (V2K) technologies explore the potential for mitigating the adverse psychological effects experienced by individuals who believe they are targets of these technologies. Many who report hearing voices or experiencing intrusive thoughts linked to V2K often suffer from anxiety, depression, and feelings of paranoia. Therapeutic approaches can range from traditional psychotherapy, where professionals help individuals process their experiences and develop coping strategies, to alternative therapies that focus on holistic healing and spiritual guidance. Understanding these interventions requires recognizing the unique psychological landscape of those affected by V2K, which often includes a blend of real distress and the influence of external narratives surrounding these technologies.

Cognitive-behavioral therapy (CBT) emerges as a potent tool for individuals grappling with the psychological ramifications of perceived V2K experiences. CBT aims to reshape negative thought patterns and beliefs, empowering individuals to challenge their perceptions of reality. This therapeutic model can help clients reframe their experiences, reducing anxiety and fostering a sense of agency. By integrating techniques that emphasize mindfulness and grounding, therapists can support individuals in distinguishing between their internal thoughts and external stimuli, which is especially crucial for those who feel overwhelmed by the implications of V2K.

Incorporating spiritual practices into therapeutic interventions offers another avenue for healing. Many individuals who believe they are channeling ethereal beings or experiencing V2K report a deep connection to spiritual realms. Therapists who respect and integrate these beliefs can facilitate a therapeutic environment that honors the individual's experience while promoting resilience. Techniques such as guided meditations, energy healing, and channeling sessions may assist in alleviating distress, allowing individuals to navigate their experiences with a sense of purpose and understanding. This approach acknowledges the multifaceted nature of human experience, blending psychological support with spiritual exploration.

The ethical considerations surrounding the use of V2K technologies also necessitate a careful examination of therapeutic interventions. Professionals in the mental health field must remain vigilant about the potential for exploitation, ensuring that their practices do not inadvertently validate harmful narratives or contribute to further stigmatization of individuals who report these experiences. Ethical frameworks in therapy encourage practitioners to maintain a compassionate stance, prioritizing clients' well-being and autonomy. This includes advocating for awareness and education about V2K technologies to demystify the experiences of affected individuals and promote informed discussions within therapeutic contexts.

As society continues to grapple with the implications of voice to skull technologies, therapeutic interventions must evolve to address the needs of those who feel impacted. This evolution will likely involve a combination of traditional psychological practices and innovative, holistic approaches that resonate with the beliefs of individuals who channel ethereal energies. By fostering a supportive environment that respects personal experiences while promoting mental health, practitioners can play a pivotal role in mitigating the psychological effects associated with V2K and paving the way for a more nuanced understanding of these complex phenomena.

Long-term Consequences

The long-term consequences of Voice to Skull (V2K) technologies extend beyond immediate concerns, delving into profound implications for individual autonomy, societal dynamics, and ethical frameworks. As these technologies gain traction in various sectors, including law enforcement and military applications, understanding their potential long-term impact becomes crucial. The ability to transmit auditory signals directly into an individual's mind poses significant questions regarding personal freedom and the right to mental privacy. Without stringent regulations, the misuse of V2K could lead to a future where individuals are manipulated or coerced through auditory intrusions, undermining the very fabric of personal agency.

Ethical considerations surrounding V2K technologies are paramount as their capabilities evolve. The potential for abuse raises red flags about consent and the morality of using such technologies on vulnerable populations. In particular, the implications for those who believe they are channeling ethereal beings or are spiritually attuned could be dire. The intersection of technology and spirituality could blur lines, leading individuals to question the authenticity of their experiences. This could foster a sense of dependency on technology for spiritual validation, eroding the intrinsic nature of personal belief systems and potentially leading to psychological distress.

The psychological effects of V2K technologies are another critical area of concern, particularly regarding long-term mental health outcomes. Continuous exposure to intrusive auditory signals could exacerbate existing mental health issues or even induce new ones, such as paranoia, anxiety, and depression. For those who already feel marginalized or targeted, the amplification of these feelings through V2K could create a cycle of victimization that is difficult to escape. Understanding these psychological ramifications is essential not only for individuals but also for communities affected by these technologies, as the collective mental well-being is at stake.

In the realm of law enforcement and surveillance, the implications of V2K technologies could alter the landscape of privacy and civil liberties. The potential for covert monitoring through auditory means raises significant concerns about the erosion of trust between citizens and authorities. If V2K becomes commonplace in policing practices, the long-term effects could include a pervasive sense of surveillance that stifles dissent and inhibits free expression. Furthermore, if these technologies are used without proper oversight, they could lead to systematic abuses that disproportionately impact marginalized groups, raising ethical and human rights issues that demand urgent attention.

Finally, the public perception of V2K technologies is shaped by a complex interplay of misinformation, fear, and intrigue. As these technologies become more visible, the narratives surrounding them often oscillate between alarming conspiracy theories and dismissive skepticism. This dichotomy complicates the discourse around V2K, making it challenging to address the genuine concerns of individuals who feel impacted by these technologies. In the absence of accurate information and transparency, the potential for societal division grows, as people may gravitate towards extreme viewpoints that further entrench their beliefs. Addressing these long-term consequences requires a multifaceted approach that promotes informed dialogue and accountability in the development and application of V2K technologies.

VII

Voice to Skull Technologies in Military Applicatio

Historical Context in Military Use

The historical context of military applications for voice to skull technologies (V2K) reveals a complex interplay between innovation, strategy, and ethical dilemmas. Initially, the military's pursuit of communication advancements can be traced back to World War II, when psychological operations began to leverage sound as a tool for propaganda and manipulation. The quest for an effective means of conveying messages directly to the minds of soldiers and adversaries alike paved the way for the exploration of non-invasive auditory technologies. This period witnessed significant investments in research focused on neuropsychology and the effects of sound on human behavior, laying the groundwork for what would eventually evolve into modern V2K systems.

As the Cold War intensified, military interest in psychological warfare surged. The United States and the Soviet Union engaged in a race not only for technological supremacy but also for the ability to control and influence the thoughts and actions of enemy combatants. This led to the development of more sophisticated audio technologies, which were theorized to induce fear, compliance, or confusion through targeted sound waves. The potential applications of such technologies were not lost on defense contractors, who began to invest heavily in research and development, envisioning a future where psychological manipulation could be executed with surgical precision.

The ethical implications of using voice to skull technologies in military contexts have been a subject of contention, particularly as the line between strategic advantage and violation of human rights becomes increasingly blurred. The potential for misuse raises significant concerns about the psychological effects on both soldiers and civilians. Incidents involving the manipulation of perceptions and thoughts could lead to profound mental health issues, including anxiety, paranoia, and disassociation. These effects extend beyond the battlefield, influencing public perception and trust in military institutions, and prompting debates on the moral responsibilities of those who wield such technologies.

Legal frameworks surrounding military use of V2K technologies have also evolved, albeit slowly. While international laws govern the conduct of war and the treatment of combatants, the rapid development of auditory manipulation technologies challenges existing legal paradigms. The lack of comprehensive regulations governing the deployment of such technologies in military operations raises critical questions about accountability and oversight. As these technologies become more integrated into military strategy, the need for robust legal safeguards that protect individual rights and promote ethical conduct is paramount.

Finally, the emergence of voice to skull technologies has not only transformed military operations but has also seeped into public consciousness, feeding into conspiracy theories and narratives surrounding government surveillance and control. The intersection of military advancement and civilian skepticism has created a fertile ground for misinformation, leading to widespread fears of manipulation and loss of autonomy. As discussions around V2K technologies continue to unfold, it is essential for both the military and the public to engage in informed dialogue about the implications, ethics, and potential futures of these powerful technologies.

Strategic Advantages

Voice to Skull (V2K) technologies present a range of strategic advantages that extend across various sectors, particularly in the realms of law enforcement, military applications, and psychological operations. These technologies enable the direct transmission of sound or messages to an individual's auditory system without the need for traditional communication devices. This capability can enhance operational efficiency, allowing for discreet communication during critical missions or investigations. The strategic use of V2K can facilitate real-time decision-making, providing law enforcement and military personnel with vital information that could be the difference between success and failure in high-stakes situations.

In the context of psychological operations, V2K technologies can serve to influence behavior and perceptions among targeted populations. By delivering specific messages or propaganda directly to individuals, it is possible to shape narratives and sway public opinion without the encumbrance of conventional media channels. This advantage poses ethical dilemmas, as the manipulation of thoughts and feelings raises questions about consent and autonomy. Yet, proponents argue that in scenarios involving national security or public safety, the benefits of controlled information dissemination may outweigh the ethical concerns.

The implications of V2K technologies extend to mental health and well-being, particularly for individuals who may experience distressing auditory phenomena. For some, the ability to modulate these experiences through technology could provide therapeutic avenues for those suffering from conditions such as PTSD or severe anxiety. However, the dual-edged sword of such capabilities highlights the necessity for careful consideration regarding the ethical use of V2K technologies. Ensuring that these tools are used to promote healing rather than harm is critical in navigating the complex landscape of mental health interventions related to voice technologies.

Public perception plays a crucial role in the strategic advantages associated with V2K technologies. Misinformation and conspiracy theories surrounding their use can lead to societal paranoia or distrust, impacting how these technologies are implemented and regulated. By fostering transparency and education about the technological underpinnings and intended applications of V2K, stakeholders can mitigate fears and enhance acceptance. A well-informed public is more likely to recognize the potential benefits of these technologies while also advocating for their ethical use, thereby creating a balanced discourse around their implementation.

Finally, the ongoing advancements in V2K technologies promise to expand their strategic applications further. As research continues to unveil the complexities of brain-computer interfaces and neurocommunication, the potential for more sophisticated interactions between technology and the human mind grows. This evolution raises important questions about legal frameworks and regulations governing the use of such technologies. Ensuring that these advancements are harnessed responsibly requires collaboration among technologists, ethicists, and lawmakers to create guidelines that protect individual rights while allowing for the beneficial uses of V2K technologies.

Ethical Concerns in Warfare

Warfare has always been a complex landscape where ethical considerations often take a backseat to strategic advantage. The advent of Voice to Skull

(V2K) technologies introduces a new layer of ethical concerns that must be critically examined, particularly as these technologies find potential applications in military operations. V2K technologies, which enable the transmission of sound directly into a person's mind, raise questions about consent, autonomy, and the psychological impact on individuals targeted by such technologies. When used in a military context, the implications extend beyond mere tactical advantages, affecting the moral fabric of warfare and the principles of just engagement.

One of the foremost ethical concerns is the issue of consent. In traditional warfare, combatants are often aware of the risks they face and can make informed choices about their involvement. However, with V2K technologies, individuals may be subjected to psychological manipulation without their knowledge or consent. This covert aspect undermines the foundational principles of warfare that emphasize the importance of informed consent and the protection of non-combatants. The potential for manipulation raises profound ethical dilemmas regarding the legitimacy of operations conducted under these conditions, as it challenges the very notion of fair play in conflict.

Moreover, the psychological effects of V2K technologies cannot be overlooked. Soldiers exposed to these technologies may experience heightened stress, paranoia, and a host of other mental health issues. The ability to implant voices or commands into an individual's mind can lead to significant trauma, not only for the direct victims but also for the military personnel who may be tasked with executing operations involving such methods. The long-term psychological ramifications on both sides of the conflict could perpetuate cycles of violence and mental distress, raising questions about the moral responsibility of those who deploy these technologies.

The implications for human rights are particularly troubling. The use of V2K technologies in warfare may violate international human rights laws that protect individuals from torture and inhumane treatment. If these technologies are used to manipulate or control individuals against their will, they could constitute a new form of psychological warfare that has far-reaching consequences.

The blurred lines between combatant and civilian can lead to widespread violations of rights, as the technologies do not discriminate between those who are actively engaged in conflict and those who are simply living in affected areas.

As society grapples with these ethical concerns, it is essential to foster a dialogue that includes diverse perspectives, particularly from those who experience the direct consequences of such technologies. Engaging with conspiracy theorists, spiritual channelers, and those who believe in ethereal communication can provide unique insights into the broader implications of V2K technologies. This dialogue is critical for developing a comprehensive understanding of the moral landscape surrounding warfare and ensuring that ethical considerations are prioritized as these technologies evolve and proliferate.

VIII

Legal Framework and Regulations Surrounding Voice

Current Laws and Policies

Current laws and policies governing Voice to Skull (V2K) technologies are complex and vary significantly across different jurisdictions. These laws are often a patchwork of existing regulations concerning surveillance, privacy, and the use of electromagnetic technologies. In many countries, legal frameworks do not specifically address V2K technologies, which can lead to significant ambiguities regarding their use in both civilian and military contexts. As these technologies evolve, there is a growing call for more explicit regulations that consider the ethical implications of using such intrusive methods.

Ethical considerations play a crucial role in the discourse surrounding V2K technologies. The potential for misuse in areas such as law enforcement and military applications raises substantial concerns. For instance, the ability to influence or manipulate an individual's thoughts or actions through auditory stimuli poses profound ethical dilemmas. These concerns are exacerbated by the lack of transparency and accountability in how these technologies are deployed. Discussions among policymakers, ethicists, and the public are increasingly focused on establishing a regulatory framework that prioritizes human rights and the protection of individual autonomy.

Psychological effects of V2K technologies cannot be understated. Numerous anecdotal accounts suggest that individuals exposed to such technologies experience a range of mental health challenges, including anxiety, paranoia, and auditory hallucinations. The implications of these psychological effects highlight the urgent need for research and dialogue surrounding V2K. Furthermore, the potential for these technologies to exacerbate existing mental health conditions raises questions about their ethical application, particularly in vulnerable populations. Public awareness and understanding of these issues are vital in pushing for more robust protections and regulations.

In the realm of law enforcement and surveillance, V2K technologies present a double-edged sword. While proponents argue that they can enhance public safety and aid in criminal investigations, critics caution against the potential for abuse and violation of civil liberties. The current legal framework often lacks the necessary safeguards to prevent misuse, leading to an environment of distrust among the public. As advancements in technology continue to outpace legislation, it is critical for lawmakers to engage with these issues proactively to ensure that civil rights are upheld and that the public is protected from potential overreach.

Public perception and misinformation regarding V2K technologies further complicate the landscape of laws and policies. Myths and conspiracy theories can distort the reality of how these technologies function and their implications for society. As a result, it becomes increasingly important for credible sources to provide accurate information and dispel unfounded fears. Engaging with communities, including conspiracy theorists and spiritual channelers, can foster a more informed dialogue about the implications, legalities, and ethical considerations of V2K technologies. By addressing misconceptions and promoting understanding, society can better navigate the intricate challenges posed by these emerging technologies.

International Regulations

International regulations regarding Voice to Skull technologies are a crucial aspect of the broader conversation surrounding the ethical and legal implications of these advanced communication methods. As the capabilities of these technologies expand, so too does the necessity for a comprehensive international framework that addresses their potential misuse. Various countries have begun to recognize the need for regulations that protect individuals from unwarranted intrusion into their mental privacy, particularly in light of the psychological effects that such technologies can have on the human psyche.

The existing legal frameworks are often fragmented and vary significantly from one jurisdiction to another. Some countries have implemented specific laws aimed at regulating the use of neurotechnologies, while others may rely on broader privacy laws that were not designed with these technologies in mind. This discrepancy creates a patchwork of regulations that can be exploited, leaving many individuals vulnerable to abuses. As the global community becomes more aware of the implications of Voice to Skull technologies, there is an urgent need for international standards that ensure consistent protection for individuals across borders.

Ethical considerations play a pivotal role in the discourse surrounding these technologies. The potential for misuse, particularly in contexts such as law enforcement and military applications, raises significant moral questions. The use of Voice to Skull technologies for surveillance or interrogation purposes can easily cross ethical boundaries, leading to violations of human rights. As awareness of these technologies grows, the demand for ethical guidelines that govern their use becomes increasingly important. Engaging in conversations about the moral implications of these technologies will help shape the development of regulations that reflect societal values and protect individual rights.

Public perception and the spread of misinformation about Voice to Skull technologies can also complicate the regulatory landscape. Many conspiracy theories and sensationalized narratives surrounding these technologies can overshadow legitimate concerns and hinder constructive dialogue. It is essential to foster informed discussions that separate fact from fiction, enabling the public to engage meaningfully in the regulatory process. By promoting transparency and education, stakeholders can contribute to a more realistic understanding of these technologies, which in turn can influence the development of sound regulations.

Finally, as technological advances continue to evolve the capabilities of Voice to Skull technologies, ongoing dialogue about international regulations must include provisions for future developments. The rapid pace of innovation can outstrip existing legal frameworks, necessitating adaptive regulations that can keep up with emerging technologies. A proactive approach to regulation can help safeguard mental health and well-being, ensuring that advancements in Voice to Skull technologies are harnessed for positive applications while mitigating risks associated with possible misuse.

Proposed Reforms

Proposed reforms surrounding Voice to Skull technologies must address the multifaceted implications of their use, particularly in areas that span ethical, psychological, and legal dimensions. As these technologies evolve, the potential for misuse raises significant concerns about individual rights and mental health. It is essential to establish a comprehensive framework that not only regulates the deployment of these technologies in law enforcement and military applications but also protects individuals from potential abuse. This framework should include strict guidelines on consent, transparency, and accountability to ensure that any implementation of Voice to Skull technologies is conducted in a manner that respects human dignity and autonomy.

Ethical considerations must be at the forefront of any proposed reforms. The transparent disclosure of the capabilities and limitations of Voice to Skull devices is crucial, especially when used in sensitive contexts such as mental health treatment or law enforcement. Ethical guidelines should mandate that individuals are informed of any potential psychological effects these technologies may have before they are subjected to them. Furthermore, a robust ethical review process should be established to evaluate new applications of Voice to Skull technologies, ensuring they align with fundamental human rights and do not infringe upon personal freedoms or mental well-being.

The psychological effects of Voice to Skull technologies cannot be overlooked. Reforms should include comprehensive mental health assessments for individuals exposed to such technologies, especially in military and surveillance contexts. Continuous monitoring and support systems must be developed to address any adverse psychological impacts that may arise. These assessments should also consider the potential for increased paranoia, anxiety, and other mental health issues associated with the perceived invasion of privacy that these technologies represent. By prioritizing mental health in the reform process, we can mitigate harmful consequences and promote a more humane approach to technology use.

Legal frameworks surrounding Voice to Skull technologies must also be re-evaluated and updated to reflect contemporary ethical standards and societal values. Current laws often lag behind technological advancements, creating gaps that could lead to abuses. Proposed reforms should call for clear legislation that defines acceptable use, implements penalties for misuse, and establishes oversight bodies to monitor compliance. Additionally, public consultations should be encouraged to gather diverse perspectives on the implications of these technologies, ensuring that the legal framework is representative of societal concerns.

Finally, addressing public perception and misinformation is critical to the successful implementation of reforms. Educational campaigns aimed at demystifying Voice to Skull technologies can help alleviate fears and misconceptions, fostering a more informed dialogue regarding their use. By engaging with communities that believe they are affected by these technologies, including spiritual and ET channelers, policymakers can gain valuable insights into the public's concerns. This engagement will not only help in crafting more effective regulations but also in promoting understanding and acceptance of technological advancements that could ultimately benefit society if used responsibly and ethically.

IX

Public Perception and Misinformation about Voice t

Media Representation

Media representation plays a crucial role in shaping public perceptions of voice to skull technologies and their implications. In an era where technology rapidly evolves, the media often serves as the primary source of information for the public. However, the portrayal of these technologies can vary significantly, leading to a spectrum of beliefs ranging from skepticism to outright acceptance. For conspiracy theorists and those who believe they are channeling ethereal beings, the media's narratives can either validate their experiences or contribute to feelings of isolation and misunderstanding. The framing of voice to skull technologies in news articles, documentaries, and online platforms influences how individuals conceptualize their experiences and understand the potential realities surrounding these technologies.

Ethical considerations are frequently highlighted in media discussions regarding voice to skull technologies. The portrayal of these tools often raises questions about consent, privacy, and the moral implications of using such technologies for surveillance or control. Media narratives that emphasize the invasive nature of these technologies can amplify fears and anxieties about their potential misuse. For those already inclined to believe in conspiracies or ethereal communications, these ethical dilemmas can reinforce their views, suggesting that unseen forces are at play. Conversely, representations that focus on the potential benefits of these technologies, such as in therapeutic or law enforcement contexts, can create a duality that complicates public understanding of their implications.

The psychological effects of voice to skull technologies are another subject of media exploration. Coverage often addresses the mental health ramifications for individuals who report experiencing these technologies, whether they perceive them as benign or malevolent. Media representations can either stigmatize those who claim to hear voices or provide a platform for their stories, affecting how society views mental health issues. For channelers and those connected to spiritual realms, the media's portrayal of mental health in relation to these technologies can impact their self-identity and the legitimacy of their experiences. This complex interplay between media representation and individual perception underscores the need for responsible reporting that considers the psychological welfare of those involved.

In the context of law enforcement and military applications, the media often presents a polarized view of voice to skull technologies. Coverage may highlight advancements in surveillance capabilities that enhance security measures, while simultaneously raising alarms about potential abuses of power. For conspiracy theorists, this dichotomy can serve to validate their beliefs about government overreach and manipulation. Reports that emphasize the dark potential of these technologies can fuel fears of a surveillance state, while those that focus on their efficacy in crime prevention may downplay ethical concerns. This duality in media representation reflects broader societal debates about the balance between safety and civil liberties, further complicating public sentiment toward these technologies.

Public perception is significantly shaped by the narratives constructed around voice to skull technologies in the media. Misinformation and sensationalist portrayals can lead to widespread misconceptions, making it challenging for individuals to discern fact from fiction. For those who believe they are channeling ethereal beings, the media's framing can either legitimize their experiences or contribute to a sense of alienation. As technological advances continue to unfold, it is essential for media outlets to provide nuanced and informed perspectives on voice to skull technologies, addressing not only their potential applications but also the ethical, psychological, and human rights implications. This comprehensive approach can foster a more informed public discourse, bridging the gap between diverse beliefs and experiences related to these controversial technologies.

The Role of Social Media

The rise of social media has fundamentally transformed the landscape in which discussions about Voice to Skull (V2K) technologies unfold. Platforms such as Twitter, Facebook, and Instagram serve not only as channels for information dissemination but also as arenas where conspiracy theories and personal experiences intersect. For individuals who believe they are channeling ethereal beings or are victims of V2K technologies, social media provides a sense of community and validation. These platforms allow users to share their narratives, often leading to a collective reinforcement of beliefs that might otherwise be dismissed by mainstream society. The ability to connect with like-minded individuals fosters a shared identity, which can intensify the conviction that they are experiencing something extraordinary or malevolent.

In the context of V2K technologies, social media plays a significant role in shaping public perception. Misinformation can spread rapidly, leading to a distorted understanding of what V2K entails and how it is purportedly used. Conspiracy theorists often leverage social media to propagate claims about government surveillance, mind control, and other nefarious applications of these technologies. This amplification of fear and uncertainty can create a feedback loop where individuals become increasingly convinced of their beliefs, regardless of the lack of empirical evidence. Furthermore, sensationalized stories can overshadow legitimate concerns regarding ethical considerations and human rights implications, ultimately hindering productive dialogue about the technology and its potential impacts.

Social media also serves as a platform for discussing the psychological effects that V2K technologies may have on individuals. Users frequently share their experiences of auditory hallucinations or intrusive thoughts, attributing these sensations to external manipulation rather than internal psychological processes. This framing can exacerbate feelings of isolation and paranoia, as individuals may begin to see themselves as targets in a larger conspiracy. The narratives constructed on social media can contribute to a sense of helplessness, as users feel they are up against a powerful and unseen force. This psychological dynamic highlights the need for a more nuanced understanding of mental health in relation to perceived technological threats.

The implications of V2K technologies in law enforcement and surveillance are also hot topics on social media, where debates about ethical boundaries and civil liberties are prevalent. Proponents argue that V2K technologies can enhance security measures, while critics raise concerns about misuse and the potential for abuse. This dichotomy often manifests in heated discussions online, where individuals express fears of a surveillance state and advocate for transparency and accountability. Social media amplifies these concerns, allowing for a diverse array of voices to contribute to the conversation, yet it can also muddle the lines between legitimate criticism and unfounded paranoia.

As technological advances continue to shape the future of V2K applications, social media will remain a crucial battleground for shaping public opinion and discourse. The impact of voice-to-skull technologies on mental health, ethical considerations, and human rights will be increasingly scrutinized in the digital realm. It is imperative for those engaged in these discussions to approach the subject with a blend of skepticism and open-mindedness, recognizing the power of social media to influence beliefs and perceptions. Engaging in thoughtful dialogue can help demystify the complexities surrounding V2K technologies and foster a more informed understanding of their implications for individuals and society as a whole.

Addressing Myths and Misconceptions

The conversation surrounding Voice to Skull (V2K) technologies is often riddled with myths and misconceptions that can distort public understanding and lead to unwarranted fears. A prevalent myth is that V2K technology is widely used for mind control or manipulation by government agencies. While the theoretical potential for such applications exists, the actual deployment of these technologies is far more limited and often subject to stringent regulations. It is crucial to differentiate between the capabilities of existing technologies and the sensationalized narratives that may arise from misunderstandings or conspiracy theories.

Another common misconception is that V2K technologies are inherently linked to extraterrestrial communication. This belief often arises from a desire to find explanations for unexplainable experiences, such as hearing voices or receiving messages that feel otherworldly. However, the science behind V2K revolves around advanced auditory technologies, primarily developed for military and law enforcement applications. These technologies are grounded in acoustic engineering and neuroscience, rather than the supernatural or extraterrestrial realms. By grounding the discussion in scientific principles, it becomes easier to address the experiences of individuals without resorting to unfounded claims.

Furthermore, the ethical considerations surrounding V2K technologies are often misunderstood. Many believe that the use of such technologies is inherently unethical, equating it with invasion of privacy or violation of human rights. While ethical implications do exist, particularly in law enforcement and surveillance contexts, it is essential to recognize the legal frameworks that govern their use. Discussions about ethics in technology must include a balanced view that considers potential benefits, such as improved communication for individuals with disabilities, alongside the risks of misuse. Engaging in informed dialogue can help demystify the intentions behind these technologies and promote a more nuanced understanding.

The psychological effects of V2K technologies often fuel misconceptions about their impact on mental health. It is not uncommon for individuals to attribute their mental health struggles or auditory hallucinations to these technologies. However, research indicates that while such experiences can feel real and distressing, they are often rooted in psychological conditions rather than external technological influences. Mental health professionals emphasize the importance of addressing these issues through therapeutic means, rather than succumbing to the belief that they are the result of targeted technological interventions. Understanding the distinction between psychological phenomena and technological influence can help individuals seek appropriate support.

Lastly, public perception of V2K technologies is often shaped by misinformation and sensationalized media portrayals. This can lead to widespread fear and misunderstanding about their actual capabilities and uses. By fostering open discussions and providing accurate information, it becomes possible to dismantle these myths and encourage a more informed perspective. Case studies and real-world applications of V2K technologies can illustrate their practical uses, as well as the safeguards that are in place to prevent abuse. Engaging with these technologies critically and thoughtfully will promote a better understanding of their implications, allowing for more constructive conversations about their role in society.

X

Technological Advances and Future Developments in

Emerging Technologies

Emerging technologies have a profound impact on how we perceive reality and interact with one another, particularly in the realm of communication. Voice to Skull technologies, which facilitate the transmission of sound directly into an individual's mind, represent one of the most controversial advancements in this field. These technologies blend neuroscience with electronic engineering, offering potential applications in various sectors, but they also raise significant ethical and psychological concerns. The implications of these technologies extend beyond mere communication; they challenge our understanding of consent, autonomy, and the very nature of human experience.

In law enforcement and surveillance, the application of Voice to Skull technologies is particularly troubling. While proponents argue that they could enhance public safety and criminal investigations, critics highlight the potential for abuse and violation of civil liberties. The ability to implant voices or commands directly into a suspect's mind raises questions about the reliability of such methods and the ethical ramifications of manipulating an individual's thoughts or actions. The balance between security and personal freedom becomes increasingly precarious, prompting a pressing need for legal frameworks that can govern the use of these technologies effectively.

The psychological effects of Voice to Skull technologies are another area of concern. Individuals who believe they are receiving messages through these means often report feelings of paranoia, anxiety, and confusion. These experiences can lead to significant mental health challenges, particularly for those already predisposed to psychological distress. The notion of being a target for such technology can exacerbate feelings of alienation and fear, impacting one's overall well-being. Moreover, the potential for misdiagnosis in mental health treatment is a dangerous consequence of the intersection between emerging technology and human psychology.

In military applications, Voice to Skull technologies offer intriguing possibilities, including psychological operations and non-lethal communication strategies. However, the ethical implications of using such technologies in warfare or conflict situations cannot be overstated. The potential for coercion and manipulation raises alarming questions about consent and the moral responsibilities of those who wield such power. As military entities explore these advanced capabilities, the need for stringent oversight and guidelines becomes crucial to prevent potential abuses and ensure that the rights of individuals are respected.

Public perception and misinformation surrounding Voice to Skull technologies further complicate the landscape. While some view these technologies with skepticism, believing them to be the stuff of conspiracy theories, others recognize their potential for misuse. This dichotomy creates a challenging environment for discourse, with many voices—both informed and uninformed—contributing to the narrative. As technological advances continue to emerge, fostering a balanced understanding of their implications, ethical considerations, and potential benefits will be critical in shaping how society navigates the complex intersections of technology, psychology, and human rights.

Potential Innovations

The exploration of potential innovations in Voice to Skull (V2K) technologies opens a complex dialogue about their implications across various fields. As advancements in neuroscience and signal processing continue to evolve, the prospect of non-invasive communication methods that directly engage the human mind becomes increasingly plausible. Future innovations may enhance the precision and clarity of auditory transmissions, potentially allowing for more nuanced and personalized communication experiences. This could lead to applications that range from therapeutic interventions for mental health conditions to educational tools that enhance learning through immersive auditory feedback.

In the realm of ethical considerations, the development of V2K technologies prompts urgent discussion about consent and autonomy. As these technologies become more sophisticated, ensuring that individuals have control over their cognitive environments becomes paramount. Innovations could focus on establishing frameworks that prioritize informed consent and the protection of personal mental space. This ethical imperative would not only safeguard individual rights but also foster public trust, which is critical for the acceptance and integration of such technologies into society.

Psychologically, the potential applications of V2K technologies could offer groundbreaking therapeutic avenues for individuals suffering from mental health issues. By facilitating communication that bypasses traditional verbal or written methods, these innovations could assist in treating conditions such as depression, anxiety, or PTSD. However, they also raise significant concerns about dependency on technological interventions and the possible blurring of the lines between self-generated thoughts and externally induced messages. A balanced approach that combines innovation with psychological safeguards will be essential in navigating these challenges.

In law enforcement and military contexts, the potential use of V2K technologies for surveillance and operational communication introduces both benefits and ethical dilemmas. Innovations could enable real-time strategic communication without the need for physical devices, allowing for discreet operations. However, this raises questions about privacy violations and the potential for misuse in authoritarian regimes. Establishing robust legal frameworks and regulations will be crucial to prevent abuses and ensure that such technologies are used responsibly and transparently.

Lastly, addressing public perception and misinformation surrounding V2K technologies will be vital as these innovations develop. The spread of conspiracy theories often stems from a lack of understanding and transparency about the science behind such technologies. Educational initiatives aimed at demystifying V2K and providing clear information on its capabilities and limitations can help bridge the gap between scientific discourse and public perception. By fostering an informed dialogue, society can better navigate the implications of these emerging technologies while addressing the fears and concerns of those who feel threatened by their potential misuse.

Ethical Considerations for the Future

Ethical considerations surrounding Voice to Skull (V2K) technologies are paramount as we move further into a future where these capabilities are increasingly intertwined with daily life. The potential applications of V2K span various sectors, including law enforcement, mental health, and military operations. However, the ethical implications of using such technologies must be evaluated thoroughly to prevent misuse and protect individual rights. A transparent dialogue about the ethical boundaries of V2K is essential for ensuring that the technology serves humanity rather than undermines it.

One significant ethical concern revolves around consent and autonomy. Individuals subjected to V2K technologies may not have given informed consent, especially in contexts like surveillance or military use. This lack of transparency raises questions about the violation of personal autonomy and the potential for coercion. For those who believe they are channeling ethereal beings or receiving messages from other dimensions, the implications of V2K extend into the realms of spirituality and personal agency, complicating the discourse on mental health and self-perception.

The psychological effects of V2K technologies are profound and multifaceted. The experience of hearing voices or receiving messages through V2K can lead to significant mental health challenges, including anxiety, paranoia, and delusions. For individuals who already face mental health struggles, the introduction of such technologies can exacerbate their conditions. It is crucial for mental health professionals and researchers to consider the impact of V2K on well-being and to develop strategies for support and intervention, ensuring that individuals are not further marginalized by emerging technologies.

In the context of law enforcement and military applications, the ethical implications become even more complex. The potential for abuse of power exists, particularly if V2K technologies are deployed without stringent regulations. The need for oversight is critical to prevent violations of human rights and to protect vulnerable populations from exploitation. Ethical frameworks must be established to govern the use of V2K in these areas, ensuring accountability and transparency while balancing the benefits of such technologies against the risks they pose to individual freedoms.

As technological advancements continue to evolve, the public perception of V2K technologies remains a crucial component of the ethical discourse. Misinformation and conspiracy theories can distort the understanding of these technologies, leading to fear and mistrust. Engaging with the public through education and open dialogue is vital for demystifying V2K and addressing concerns. By fostering an environment where ethical considerations are prioritized, we can work towards a future where technology enhances human experience rather than diminishes it.

XI

Case Studies of Alleged Voice to Skull Technology

Notable Incidents

Notable incidents surrounding Voice to Skull technologies offer a compelling glimpse into the complexities and controversies associated with these systems. One of the most frequently cited cases occurred in the early 2000s when numerous individuals in various countries reported experiencing auditory hallucinations that they believed were the result of targeted harassment using advanced sonic technology. Many of these individuals claimed that they could hear voices that were not present in the physical environment, leading to significant distress and challenges in their daily lives. These reports generated a wave of interest among conspiracy theorists, who argued that such technologies were being used covertly by government agencies for mind control or psychological manipulation.

Another significant incident involved allegations from military personnel who claimed to have been subjected to experimental applications of Voice to Skull technologies during training exercises. These individuals reported experiencing disembodied voices that instructed them to perform specific actions or provided guidance during high-stress scenarios. The accounts of these veterans raised ethical questions about the use of such technology in military settings, particularly regarding informed consent and the psychological impact on service members. The blurred lines between training simulations and real psychological effects present a troubling aspect of Voice to Skull technology applications.

In 2015, a case in the United States drew public attention when a woman filed a lawsuit against a government agency, alleging that she had been a victim of Voice to Skull technology harassment. Her claims included assertions that she was being targeted by a covert program designed to manipulate her thoughts and actions through auditory means. This incident not only highlighted the personal impact of these technologies but also sparked a broader discussion about legal protections for individuals who believe they are victims of such intrusive practices. The case exemplifies the challenges faced by those seeking to navigate the intersection of technology, personal rights, and mental health.

Public perception of Voice to Skull technologies has often been shaped by sensationalist media coverage and misinformation. Many people remain skeptical or misinformed about the capabilities and intentions behind these technologies. This skepticism can lead to dismissiveness regarding the genuine experiences reported by individuals who believe they have been affected. The challenge lies in fostering an informed discourse that distinguishes between legitimate concerns over human rights implications and unfounded conspiracy theories. Misunderstandings can further isolate those who experience distressing auditory phenomena, making it essential for advocates to promote awareness and understanding of the psychological and ethical dimensions of these technologies.

As technology continues to evolve, the potential for both beneficial and harmful applications of Voice to Skull technologies remains a pressing issue. Recent advancements have led to more sophisticated systems capable of precise targeting and enhanced sound delivery, raising concerns about their misuse in various contexts. The ethical landscape surrounding these technologies is still developing, and ongoing discussions will likely shape future regulations and societal perceptions. Understanding the notable incidents related to Voice to Skull technologies not only sheds light on the broader implications but also underscores the need for vigilance in protecting individual rights and mental well-being in an increasingly technologically driven world.

Analysis of Claims

The analysis of claims regarding Voice to Skull Technologies (V2K) reveals a complex interplay of technological advancements, ethical considerations, and psychological implications. Proponents of V2K assert that these technologies allow for the transmission of auditory signals directly into an individual's mind, bypassing traditional auditory pathways. Such claims raise significant questions about the validity of the technology itself and the extent to which it can influence thoughts and behaviors. To understand these claims, it is crucial to differentiate between scientifically substantiated evidence and anecdotal reports that often characterize discussions surrounding V2K.

Ethical considerations are paramount in the discourse on V2K technologies. The potential for misuse in surveillance and law enforcement settings poses serious moral dilemmas. If these technologies can indeed facilitate remote communication or influence decision-making without consent, they challenge fundamental human rights. The analysis must consider the implications of such capabilities, particularly in terms of privacy, autonomy, and the potential for abuse by state or corporate entities. Ethical frameworks must be established to govern the development and application of V2K, ensuring that technological advancements do not infringe upon individual freedoms.

The psychological effects of V2K must also be scrutinized, especially given the claims of individuals who believe they are receiving messages from ethereal beings or higher consciousness. Reports of auditory hallucinations and perceived communication with external entities can lead to significant mental health challenges. Understanding the mechanisms behind these experiences is critical for both supporters and skeptics of V2K. Psychological evaluations and case studies can provide insight into how belief in these technologies can exacerbate existing mental health conditions or create new ones, blurring the line between reality and perception.

In law enforcement and military applications, V2K technologies present a unique set of challenges and opportunities. The use of such technologies for interrogation, crowd control, or psychological operations raises critical questions about legality and morality. The potential for coercive psychological influence poses risks not only to targeted individuals but also to the integrity of justice systems. Analyzing case studies where V2K has allegedly been employed can shed light on the ramifications of these practices, revealing the balance between security measures and civil liberties.

Finally, public perception and misinformation play a significant role in shaping the discourse around Voice to Skull Technologies. The proliferation of conspiracy theories surrounding V2K can hinder rational debate and understanding of the actual capabilities and limitations of these technologies. As developments continue in the field, it is essential to foster a well-informed public dialogue that distinguishes between legitimate scientific inquiry and sensationalist claims. This understanding will be crucial in navigating the future of V2K technologies and their impact on society, mental health, and human rights.

Implications of Case Studies

The exploration of Voice to Skull technologies through various case studies reveals profound implications that extend across ethical, psychological, and legal dimensions. In examining specific instances where these technologies have allegedly been deployed, we gain insights into their potential to influence human behavior and thought processes. The implications of these case studies suggest not only the feasibility of such technologies but also the ethical dilemmas they pose, raising questions about consent, autonomy, and the right to mental privacy. As individuals report experiences of auditory hallucinations or perceived communication with external entities, it becomes essential to scrutinize the validity and authenticity of these experiences within the context of technological capabilities.

Ethical considerations emerge as a significant theme within the narrative of Voice to Skull technologies. The case studies highlight scenarios where individuals have claimed to be subjected to non-consensual experimentation, leading to debates on the morality of using such technologies without explicit permission. These narratives challenge the established norms regarding human rights, particularly in relation to mental integrity and the right to an untainted thought process. The implications extend beyond individual cases, prompting a broader discussion on the governance of emerging technologies and the need for strict ethical guidelines to safeguard against potential abuses.

The psychological effects of Voice to Skull technologies, as illustrated by case studies, reveal a spectrum of responses among individuals who believe they are experiencing these phenomena. Reports of anxiety, paranoia, and altered states of consciousness are prevalent, suggesting a significant impact on mental health and well-being. These technologies can potentially exacerbate existing mental health issues or create new psychological challenges for individuals, especially those who may already be vulnerable. The implications of these findings call for a comprehensive assessment of the interactions between technology and mental health, emphasizing the necessity for support systems for those affected.

In the realm of law enforcement and surveillance, case studies demonstrate the potential application of Voice to Skull technologies as tools for interrogation and monitoring. While proponents may argue for their efficacy in enhancing security measures, the implications for civil liberties and human rights are troubling. The potential for misuse by authorities raises critical questions about accountability and the ethical limits of surveillance technologies. These narratives serve as cautionary tales, urging society to tread carefully when integrating such technologies into public safety frameworks, lest we compromise fundamental freedoms.

Finally, the public perception of Voice to Skull technologies, shaped by case studies and anecdotal evidence, reflects a complex landscape of misinformation and skepticism. Individuals who assert their experiences often face disbelief, which can further isolate them and complicate their mental health journeys. The implications of this dynamic highlight the urgent need for education and awareness surrounding these technologies, fostering a more informed discourse that separates fact from fiction. As technological advances continue to unfold, the interplay between public perception, ethical considerations, and the rights of individuals will be crucial in shaping the future landscape of Voice to Skull technologies.

XII

Voice to Skull Technologies and Human Rights Impli

Right to Privacy

The right to privacy is a fundamental aspect of human dignity and autonomy, particularly in an age where technologies like Voice to Skull (V2K) are becoming increasingly sophisticated and pervasive. This technology, which allows for the transmission of auditory messages directly into a person's mind, raises significant ethical and legal questions regarding individual rights. As conspiracy theorists and spiritual channelers often highlight, the implications of such technologies extend far beyond mere communication; they touch upon the essence of personal freedom and the sanctity of one's inner thoughts. In an environment where privacy can be breached without consent, individuals may find their mental space invaded, leading to a profound sense of vulnerability.

The ethical considerations surrounding V2K technologies are complex and multifaceted. Many proponents of privacy argue that the ability to transmit thoughts or commands directly into a person's consciousness constitutes a severe infringement on personal autonomy. This technology challenges the very concept of consent, as individuals may be targeted without their knowledge or agreement. The potential for misuse is staggering; it could facilitate manipulation, coercion, or even psychological torture. For those who believe they are channeling ethereal beings, the line between genuine spiritual experiences and intrusive technological influence becomes increasingly blurred, prompting concerns about the authenticity of their experiences and the integrity of their mental landscapes.

The psychological effects of V2K technologies can be profound and damaging. Individuals exposed to unsolicited auditory messages may experience heightened anxiety, paranoia, and a sense of loss of control over their thoughts. This can lead to a deterioration of mental health, complicating existing psychological conditions or even inducing new ones. The mind, already a complex interplay of thoughts and emotions, faces challenges when external voices intrude upon its sanctity. For those who are spiritually inclined, the experience might also lead to confusion about their spiritual practices, as distinguishing between internal thoughts and externally imposed narratives becomes increasingly difficult.

In the context of law enforcement and surveillance, the use of V2K technologies raises critical human rights concerns. While proponents may argue that such tools could enhance public safety, the potential for abuse is significant. The ability to implant thoughts or directives could easily lead to wrongful accusations or manipulative interrogations. Furthermore, the lack of a robust legal framework governing these technologies leaves individuals vulnerable to exploitation by both state and non-state actors. The ramifications of this practice reach into the core of democratic values, undermining trust in institutions that are meant to protect individual rights and freedoms.

The public perception of V2K technologies is often colored by misinformation and fear, which can further complicate discussions around privacy and ethics. Many people may dismiss these technologies as mere conspiracy theories, while others may genuinely believe they are victims of such experiences. This division creates a challenging landscape for advocacy and regulation, as the voices of those who suffer from the negative impacts of V2K may be drowned out by skepticism. A comprehensive understanding of the implications of these technologies is crucial not only for those who feel targeted but also for society at large, as the ongoing developments in V2K reveal the need for clearer legal protections and ethical guidelines that uphold the right to privacy in an increasingly interconnected world.

Freedom of Expression

Freedom of expression is a fundamental human right that allows individuals to articulate their thoughts, beliefs, and emotions. In the context of Voice to Skull (V2K) technologies, this principle becomes particularly complex. V2K technologies, which can transmit auditory messages directly into a person's mind, raise important questions about the limits of free speech and the potential for abuse. The ability to influence or manipulate an individual's thoughts through such technologies presents a chilling scenario where the essence of personal autonomy and expression could be compromised. This intersection of technology and personal rights necessitates a critical examination of how freedom of expression is upheld or threatened by these advancements.

The ethical considerations surrounding V2K technologies are profound. On one hand, the potential for these technologies to facilitate communication for individuals with disabilities can be seen as an advancement in human expression. On the other, the risk of misuse, such as coercive persuasion or psychological manipulation, poses a significant threat to the integrity of personal expression. The ability to project thoughts or voices into a person's mind can lead to scenarios where individuals are compelled to express ideas or opinions that are not their own, raising ethical dilemmas about consent, autonomy, and the authenticity of one's voice.

Psychologically, the impact of V2K technologies on mental health cannot be overlooked. For individuals who may already be vulnerable, the experience of hearing voices or receiving messages from an external source could exacerbate feelings of paranoia, anxiety, or distress. The implications for mental well-being are particularly concerning within communities that are already marginalized or skeptical of mainstream narratives. The potential for these technologies to disrupt an individual's inner dialogue or sense of self-heightens the urgency for discussions about how freedom of expression can coexist with the risks posed by these intrusive technologies.

In the realms of law enforcement and military applications, the deployment of V2K technologies raises significant legal and ethical questions. While proponents may argue that such tools could enhance national security or public safety, the potential for abuse and the infringement on civil liberties is a pressing concern. Surveillance practices that incorporate V2K technologies could lead to a society where free expression is stifled, as individuals may self-censor out of fear of being monitored or manipulated. The balance between security and freedom of expression must be carefully navigated to avoid creating an environment of distrust and repression.

Finally, the public perception of V2K technologies is shaped by misinformation and sensationalism, further complicating the discourse on freedom of expression. As conspiracy theories proliferate, the line between credible concerns and unfounded fears blurs, leading to a polarized understanding of these technologies. Educating the public about the realities and implications of V2K technologies is essential for a balanced discourse. Only through informed dialogue can society hope to protect the sanctity of personal expression while addressing the ethical and psychological ramifications of these powerful technologies.

Advocacy and Activism

Advocacy and activism surrounding Voice to Skull technologies have emerged as critical components in the discourse about their ethical implications and potential impacts on society. As awareness of these technologies grows, so does the concern over their misuse, particularly in contexts that infringe on individual rights and freedoms. Groups advocating for transparency and accountability seek to shed light on the potential for abuse, especially regarding military and law enforcement applications. By mobilizing communities and utilizing various platforms, these activists aim to challenge the narrative surrounding these technologies, raising essential questions about their ethical deployment and the consequences for human rights.

One of the primary concerns of activists is the psychological impact of Voice to Skull technologies on individuals, especially those who may already be vulnerable. Reports have surfaced of individuals experiencing distressing auditory phenomena, leading to severe mental health challenges. Advocacy groups work to bring attention to these issues, emphasizing the need for comprehensive mental health support for those affected. Additionally, they push for further research into the psychological effects of such technologies, highlighting the importance of understanding how they can alter a person's perception of reality and their overall well-being.

In the realm of law enforcement and surveillance, the implications of Voice to Skull technologies present a complex ethical landscape. Activists argue that the potential for these technologies to be used as tools for coercion or manipulation poses a significant threat to civil liberties. There is growing concern that, without adequate regulations, law enforcement agencies could exploit these technologies to infringe upon personal freedoms and privacy. Advocacy efforts focus on the necessity of legal frameworks that govern the use of such technologies, ensuring they are not deployed in ways that violate human rights or undermine trust in public institutions.

Public perception of Voice to Skull technologies is often clouded by misinformation and sensationalism, which can hinder meaningful dialogue about their implications. Advocacy groups aim to combat these misconceptions by providing accurate information and fostering open discussions about the realities of these technologies. By engaging with communities and utilizing social media campaigns, activists strive to create a more informed public that can critically assess the potential dangers and ethical dilemmas surrounding Voice to Skull technologies. This grassroots approach seeks to empower individuals to advocate for their rights and for the responsible use of technology.

As technological advancements continue to evolve, so too does the landscape of advocacy and activism related to Voice to Skull technologies. Activists are increasingly harnessing digital tools and platforms to raise awareness and mobilize support. The future of advocacy will likely involve collaboration between various stakeholders, including mental health professionals, legal experts, and technologists. This collective effort is essential in shaping policies and regulations that prioritize ethical considerations and protect individuals from the adverse effects of these technologies. Ultimately, the goal of advocacy in this area is not only to address immediate concerns but also to foster a broader conversation about the ethical use of emerging technologies in society.

XIII

Conclusion and Future Perspectives

Summary of Key Findings

This subchapter summarizes the key findings from the exploration of Voice to Skull technologies, shedding light on their multifaceted implications. One of the primary conclusions drawn is the profound ethical considerations surrounding these technologies. The ability to transmit auditory information directly into an individual's mind raises significant moral questions about consent, autonomy, and the potential for misuse. As the line between technology and personal agency becomes increasingly blurred, the exploration of these ethical dimensions is crucial for both developers and users of such technologies.

The psychological effects of Voice to Skull technologies present another area of concern. Research indicates that exposure to these technologies can lead to a range of psychological responses, including anxiety, paranoia, and altered states of consciousness. The potential for these technologies to be used as a tool for manipulation or control demands a thorough understanding of their impact on mental health and well-being. It is essential to consider how these effects can vary across different populations, particularly for individuals already experiencing psychological distress.

In the context of law enforcement and surveillance, Voice to Skull technologies emerge as a double-edged sword. While they offer innovative methods for communication and information dissemination, their application raises significant concerns regarding civil liberties and privacy. The capacity for authorities to deploy these technologies in a covert manner could lead to abuses of power and unwarranted intrusions into personal lives. A nuanced discourse on the balance between security and individual rights is necessary to navigate this complex landscape.

Military applications of Voice to Skull technologies further complicate the narrative. Their potential to enhance communication in combat scenarios presents strategic advantages, yet the implications for psychological warfare and the ethical treatment of soldiers must be examined. The military's interest in such technologies underscores the need for clear regulations and oversight to prevent potential exploitation or harmful consequences for service members.

Lastly, public perception and misinformation surrounding Voice to Skull technologies pose significant challenges. Misunderstandings and conspiracy theories can obscure the factual basis of these technologies and hinder constructive dialogue. As advancements continue to emerge, fostering an informed public discourse is essential. Education and transparency about the capabilities and limitations of Voice to Skull technologies will be vital in addressing fears and ensuring that developments in this field align with ethical standards and human rights principles.

Call to Action

As we delve deeper into the intricate web of Voice to Skull technologies, it becomes crucial for you, the audience, to engage actively with the information presented. These technologies, designed to transmit audible messages directly into the human brain, pose profound ethical and psychological implications that warrant your attention. The potential for misuse in various sectors, including law enforcement and military applications, raises critical questions about the rights of individuals and the balance between security and personal freedom. Understanding these dimensions is not just an intellectual exercise; it is a call to awareness and action.

The ethical considerations surrounding Voice to Skull technologies must not be overlooked. As conspiracy theorists, spiritual seekers, and those who believe in channeling ethereal beings, you are uniquely positioned to question the motives behind the development and deployment of such technologies. By recognizing the potential for abuse, you can advocate for transparency and accountability in their application. Engaging in discussions and forums can amplify your voice, ensuring that the ethical dimensions of these technologies are at the forefront of public consciousness.

Additionally, the psychological effects of Voice to Skull technologies on mental health and well-being are areas ripe for exploration. The potential for these technologies to exacerbate feelings of paranoia or anxiety cannot be dismissed. By sharing personal experiences and insights, you contribute to a collective understanding of how such technologies impact individuals on a psychological level. This shared knowledge can foster support networks and promote resilience among those who feel affected, encouraging a community-driven approach to mental health in the context of technological advancements.

Legal frameworks and regulations surrounding Voice to Skull technologies are still in their infancy. As believers in the implications of such technologies, your advocacy can shape future policies. Engaging with lawmakers and participating in public discourse will help ensure that any legislation enacted prioritizes human rights and safeguards against potential abuses. It is essential to voice your concerns and aspirations for a legal system that recognizes the profound implications of these technologies on society.

Lastly, public perception and misinformation about Voice to Skull technologies must be addressed. As individuals who embrace the exploration of hidden truths, you play a pivotal role in disseminating accurate information and combating sensationalist narratives. By sharing research, case studies, and personal testimonies, you can foster informed discussions that demystify these technologies. Empowering others with knowledge will not only enhance understanding but also inspire a collective call to action, urging society to confront and navigate the complexities inherent in Voice to Skull technologies.

Future Research Directions

Future research directions in the realm of Voice to Skull (V2K) technologies should focus on understanding the implications of these systems on various aspects of society, particularly for those who perceive themselves as recipients of ethereal communications. Research can delve deeper into the psychological effects experienced by individuals who believe they are being targeted by V2K technologies. This includes exploring the intersection of belief systems and mental health, where study findings could provide insights into the experiences of perceived victims, potentially leading to a greater understanding of the psychological ramifications associated with these technologies. By investigating the narratives of individuals who claim to hear voices through V2K, researchers can uncover patterns that inform both psychological support and technological ethics.

Ethical considerations surrounding V2K technologies warrant thorough examination, particularly in the context of consent and human rights. Future research should aim to develop frameworks that address the ethical dilemmas posed by the use of such technologies in surveillance and law enforcement. By engaging with communities that feel targeted, researchers can create dialogues that prioritize ethical governance and develop guidelines for the responsible use of V2K technologies. This approach would also encompass the implications for marginalized groups, ensuring that research takes into account diverse perspectives and experiences.

The impact of V2K technologies on mental health and well-being remains a critical area for future inquiry. As the stigma surrounding mental health continues to diminish, it is essential to understand how individuals who believe they are experiencing V2K phenomena can be supported. Research could focus on the therapeutic approaches that can assist these individuals, potentially integrating traditional psychological practices with alternative spiritual or metaphysical perspectives. This interdisciplinary approach may lead to more comprehensive support mechanisms and a reduction in the distress associated with perceived V2K experiences.

Technological advances in V2K systems present both opportunities and challenges. Future research should investigate the potential developments in these technologies, considering how they might evolve and the implications of these advancements for society. This includes understanding how emerging technologies could exacerbate existing issues related to privacy, consent, and surveillance. Additionally, studies could explore innovative ways to counteract the negative impacts of V2K technologies and enhance public understanding to combat misinformation and fear surrounding these systems.

Lastly, examining legal frameworks and regulations concerning V2K technologies is essential for shaping future discourse. Research efforts should focus on identifying gaps in current legislation and advocating for policies that protect individuals from potential abuses. This includes a critical analysis of case studies that highlight the use of V2K technologies in military applications and their implications for civil rights. By fostering discussions on legal accountability and human rights, researchers can contribute to a more informed public perception and encourage a balanced narrative that incorporates the voices of those who feel affected by these technologies.